NEURAL ACTIVITY POTENTIAL IN DEAD BONES THEORY.

Written by: Sheila Ber.

INTRODUCTION:

I majored in Science in 1991 from the University of Toronto. Physics and Chemistry were especially, my favourite subjects. I'm affiliated with The Chemical Institute of Canada.

I wrote this theory, as a result of my curiosity for the supernatural phenomena.
My desire was to analyze the subject of renewed neural activity in dead bones from a perspective that is simple, and fundamentally scientific.

I have no references, as I did not refer to any literature. It is solely my own idea.

I hope that you'll find my theory interesting.

Sheila Ber
Publisher

e-mail address: blueiris0000@yahoo.com

MY SCIENTIFIC THEORY

Fully decomposed dead bodies, leave only the bones lying underground. Since the bones of the living have nerve cells. The bones of a dead body, I believe still probably retain their nerve cells, but they are in a dehydrated state. When they get hydrated however, by getting in contact with water seeping underground, they may have the capacity to function at some minimum level, which is of course at a different capacity, versus nerve cells in a living body.

To activate nerve cells, they require minerals, Oxygen, optimum temperature, and electricity. The first two variables are found in water, and of course the fourth one, electricity, is generated by the great cosmic electromagnetic and/or electrostatic force. When these variables are combined, they produce a synergistic action, which may at an optimal temperature, spark life in dead bones.

Electromagnetic fields have generally a profound effect on our nervous system, when we are alive, my understanding is, that they can also have an effect, at a different level, on nerve cells of a dead body lying underground, whether hydrated or dehydrated.

The effect is greater however when a body, or even just the bones of a dead body, are in a re-hydrated state, again due to contact with mostly rain water seeping underground.

Electromagnetic and electrostatic fields exist nearly everywhere, and they gain more strength in saturated air, with a relative humidity that is high. As we already know, electricity travels faster in water, because water is highly conductive.

Every dead body has nerve cells, but of a different capacity, and therefore, will be affected accordingly, or perhaps will not be affected at all.

Those who have higher nerve cells capacity may, with the influence of the external strong electromagnetic and electrostatic forces, have the ability to transmit electromagnetic signals or non verbal messages, from the underground and into the atmosphere.

Electromagnetic and electrostatic forces may therefore, have the ability to spark a renewed neural activity in dead bones, given the right conditions as mentioned above. The right conditions can serve as catalysts for electrochemical reactions that may reactivate nerve cells.

Electromagnetic signals may be interpreted in several forms. Some of the familiar ones are:
Non-verbal communication i.e. messages, or, ghosts appearing as visible light faded shapes, that are formed by aura energy fields.

The sun and the moon exert forces and pressure on earth, and its objects. The most influential ones are the gravitational, electrostatic, and the electromagnetic force.

When the moon is in its full phase, it exerts stronger gravitational, electrostatic, and electromagnetic forces on earth, and everything on it and within it.

The combined forces of the moon, and the sun, are especially strong at the full moon phase, because the moon is situated directly between the Earth and Sun.

Furthermore, nerve cells in the bones of a skull, have probably retained historical data of a dead person's past life. If energized and awakened by the factors as discussed above, and therefore have the capacity to release electromagnetic signals, such signals may be either weak or strong.

The signals from a dead person, would probably be received and felt more easily, by the closest relative/s or friend/s of the dead person that's sending the signals. Relatives and friends may be the ones more familiar with the content of such signals.

I therefore deduced that the above forces which are relatively very strong during full moon phase period, may have stronger reviving effect on dead bones' nerve cells. These forces generate stronger energy fields affecting not only the living, but the dead as well.

CONCLUSION:

Strong electromagnetic and electrostatic forces might enable re-hydrated nerve cells in dead bones, to transmit electromagnetic signals that travel through the underground to the atmosphere.
These signals may contain information such as feelings of pain, joy, anger, or even revenge.

However, I must point out lunar and solar forces do not have sufficient power to bring the dead back to life, to form a fully functioning human body.

Again, these signals could be received, felt, and also interpreted by the living entities, who are maybe more familiar with the content of such signals. Usually, it would be the relatives or friends of the dead person that transmit those signals.

*Two research groups reported in 1 issue of *Science* Journal, that bone marrow cells can travel to the brain and turn into nerve cells.

The bone marrow cells have the capacity to change into nerve cells. They appear to be making proteins that nerve cells make.
*The bone marrow does have a rich blood and nerve supply.

Feelings are nothing but electrical impulses felt by you in your brain. Feelings are usually often triggered by neuro-chemical agents.

Now, who says dead bones can't have feelings too?

*Similar to the evolution theory, unfortunately my theory would be very difficult to prove, and I accept that.

Nerve cells diagram:

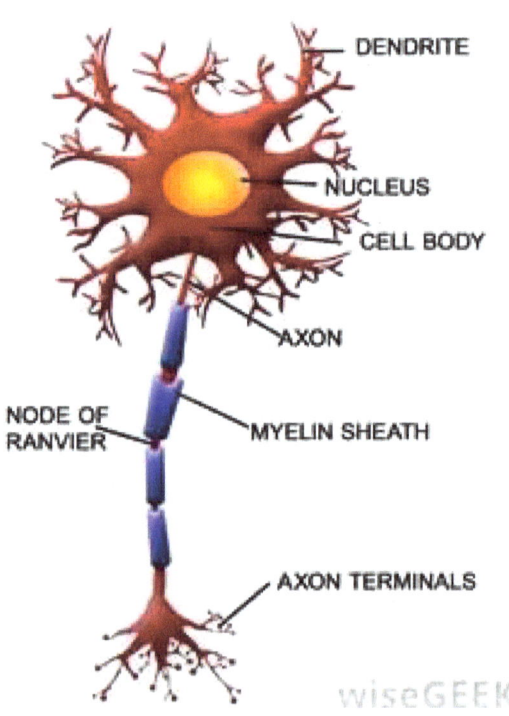

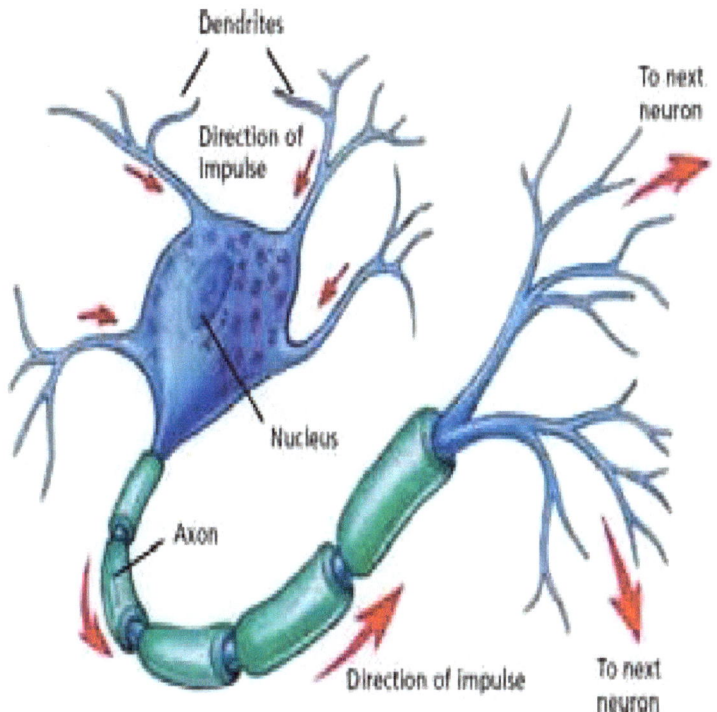

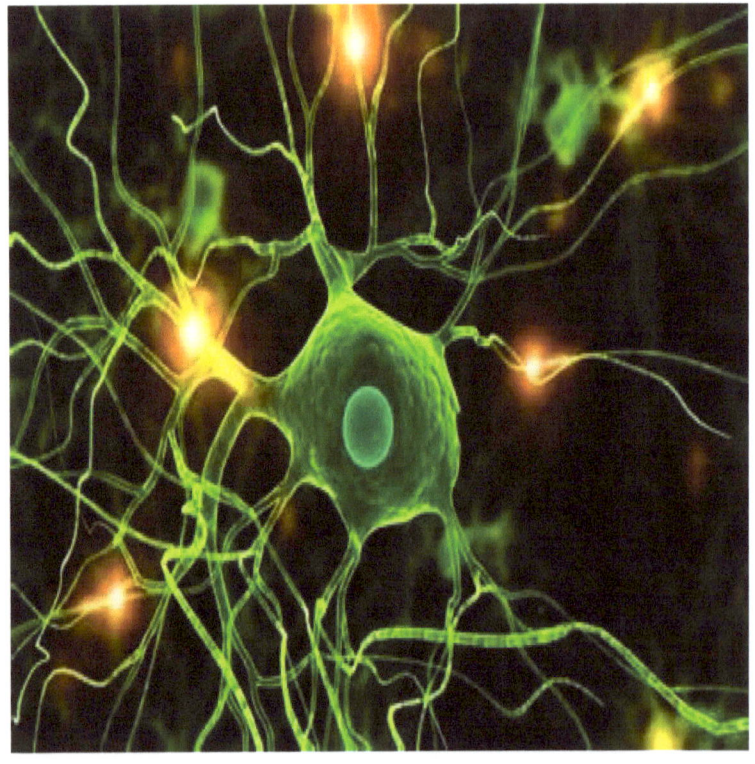

High resolution image of nerve cells dyed in green.

Keywords:

Nerve cells, electromagnetic signals, electrostatic force, electromagnetic force, lunar and solar forces, spark of life, energy fields auras, neural activity, ghosts, spirits, supernatural, dead bones, electrochemical reactions.

References:

1) Foster, Russell G.; Roenneberg, Till (2008). "Human Responses to the Geophysical Daily, Annual and Lunar Cycles". Current Biology 18 (17): R784–R794. doi:10.1016/j.cub.2008.07.003. ISSN 0960-9822. PMID 18786384.
2) Carroll, Robert Todd (12 August 2011). "Full Moon and Lunar Effects". The Skeptic's Dictionary. Retrieved 22 October 2011.

Disclaimer.

<u>Biography</u>

I majored in Science in 1991 from the University of Toronto. Physics and Chemistry were especially, my favourite subjects.

I worked in Microbiology and Chemistry, for about 12 years, in the Pharmaceutical, cosmetics, and toiletry industries.
I was involved in Research & Development, analysis and in formulations of large variety of products.

I am an unconventional person, at the same time I like things straight, simple, and uncomplicated.

I have a tendency to analyze everything almost to death. Of course this habit has its positives but also some negative implications.
I like helping people. I view people, things, situations, from different perspectives, and try to remain neutral.

Our present digital world is a somewhat intimidating, but is rather promising at the same time. It is best to exercise the right balance in our lives.

SHEILA (SHULLA) BER

GHOSTS: FACT OR FICTION?

Now in: 1. www.amazon.com

2. www.kobobooks.com

www.ingramcontent.com/pod-product-compliance
Lightning Source LLC
Chambersburg PA
CBHW041624180526
45159CB00002BC/997